Teaching Nursing in the Clinical Setting
A Practical Guide to Clinical Instruction

by Helen Spustek O'Shea, RN, PhD
Professor Emerita
Nell Hodgson Woodruff School of Nursing
Emory University
Atlanta, Georgia

©2008 Helen Spustek O'Shea

All rights reserved. No part of this publication may be reproduced or distributed in any form or by any means without the written permission of the publisher.

ISBN-13: 978-0-6151-9526-1

Lullwater Press
Atlanta, GA
http://www.lullwaterpress.com/

TABLE OF CONTENTS

Section 1	The Role of the Clinical Instructor	1
Section 2	Understanding the Clinical Course as a Curriculum Component	5
Section 3	Acquainting Yourself with the Learners	13
Section 4	Getting Oriented to the Clinical Setting	19
Section 5	Determining the Clinical Assignment	43
Section 6	Conducting the Clinical Lab	53
Section 7	Evaluating Clinical Performance	63
Section 8	Legal Considerations of Clinical Teaching	71

ACKNOWLEDGEMENTS

This book has been under construction for a long a time and I have used it as a handout in the nursing education courses I taught over many years at the Nell Hodgson Woodruff School of Nursing at Emory University. I would like to offer a grateful "thank you" to all of the masters, post masters, and doctoral students who took the time to respond to my requests for feedback about the book in all of its developmental stages. Their comments and suggestions about content and format have been most helpful. Thanks also go to my faculty colleague Corrine Abraham, RN, MS who so graciously agreed to serve as the clinical instructor in the photographs that begin each chapter.

Special thanks and recognition go to my husband, Donald C. O'Shea, PhD, Professor Emeritus at the Georgia Institute of Technology. He provided ongoing support and encouragement over the long haul and spent many hours working with formatting the text for publication. In addition he is responsible for its overall design, photographs, and illustrations. And, yes, we are still happily married and good friends.

Helen S. O'Shea
Atlanta, Georgia
January, 2008

PREFACE

This book has been in the making for over 20 years. It started out as a set of handouts to graduate students enrolled in a teaching major at the School of Nursing at Emory University. It has grown in size and scope over the years and became a draft manuscript that I have distributed to graduate students, continuing education students, doctoral students, and most recently to post-masters students in a teaching certificate program. It is based on 32 years of my own experience as a clinical teacher of undergraduate nursing students in adult medical-surgical settings in major teaching hospitals. It has been enriched through many shared experiences with my faculty colleagues who are also actively engaged in clinical teaching.

It is a practical and pragmatic discussion of how to help nursing students make the transition from terrified lay people to confident and competent novice professional nurses. By the same token the guidelines, suggestions, and strategies offered in this book should help an experienced nurse make the transition from nervous novice to confident and competent clinical instructor. Expertise in teaching will develop over time and with experience just as it does in the direct care practice role.

I have deliberately decided to not include references from the professional literature. My rationale is that the field of teaching, like the field of clinical practice, is undergoing change. The reader would be far better served to go to the literature to find current answers to questions that arise about best practice or about best teaching strategies than to depend solely on articles listed by the author that were current at the time of writing but may no longer be current at the time of reading.

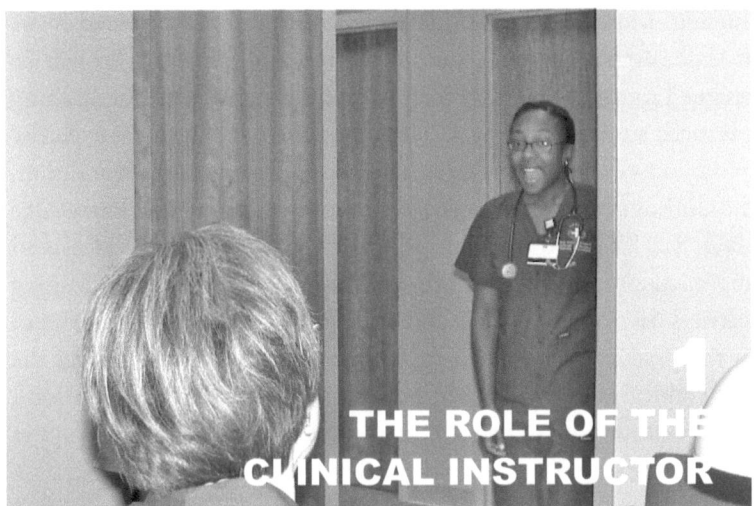

1
THE ROLE OF THE CLINICAL INSTRUCTOR

Nurse educators who teach students in clinical practice settings are privileged people. They have the best of both the academic and the practice worlds. It is in the clinical practice setting that knowledge becomes meaningful to students and it is the clinical instructor who chooses the experiences that allow the students to apply their knowledge. The clinical instructor has many opportunities to make learning relevant for students and to contribute to high quality nursing care for patients. Although it is more in vogue to use the word, "client," when writing about persons in need of nursing care I have chosen deliberately to use the word patient because I believe that, for the clinical instructor, the client is the student nurse.

If one accepts the concept of the student as the client, then the role of the instructor is to work with the student to assess learning needs, plan and implement learning experiences, evaluate results, and, if necessary, revise the plans and try again. The instructor also serves as student advocate, encourages independence, and makes appropriate referrals when someone else could better meet the student's needs.

The term clinical instructor carries with it certain assumptions and expectations not the least of which is that the person who functions as a clinical instructor is a competent practitioner of nursing. More specifically, the clinical instructor is expected to be at least competent in the care of the types of patients who will be assigned to the students. If the instructor is responsible for teaching advanced students, he or she should be able to demonstrate expertise in the care of the patients to be assigned to the students. In addition, the clinical instructor is expected to possess fundamental knowledge about teaching and learning principles, be open to new ideas and approaches to nursing practice, and have a commitment to helping learners develop confidence and at least beginning level competence in the practice of nursing. The clinical instructor is bound by the same ethical code as all other professional nurses, and the instructor has a responsibility to help students recognize the role ethics plays in decision making related to patient care.

At times, the role of the nurse educator in the clinical setting can be ambiguous. While instructors may embrace the concept of the student as a client, they cannot ignore the responsibility they have as professional nurses to assure that the patient suffers no harm as a result of the actions of an inexperienced student. It is a dilemma that the clinical instructor faces frequently and it has no predetermined solutions.

In some cases the decision to switch from the teacher to the nurse role has to be made quickly in order to avert potential harm or to reduce the risk of harm. If the instructor finds that it is necessary to assume the nurse role with great frequency it would be advisable to examine the situation carefully. Sometimes the patient is too complex for the student to manage independently.

There may be reasonable doubt that a particular student is capable of providing care that other students in the same clinical group would be able to perform with no or minimal assistance. The instructor may be expecting a level of speed, competence, or expertise that is unrealistic for the level of the learner. The instructor may be so eager to be a role model for professional practice that the student is not being given the additional time learners usually need to think through each step of the care process. Although it is likely that the instructor can perform almost any procedure more quickly, more efficiently, and more skillfully than the student, and although the instructor may believe that demonstrating his or her expertise by performing a procedure is a valid form of role modeling, it is worth remembering that there can be a fine line between being a role model and being a show off. In fact, the instructor's expertise can have the effect of intimidating rather than encouraging the student.

Despite the seemingly endless responsibilities, clinical teaching can be the most rewarding part of being a nurse educator. It is like having your cake and eating it too. You get to choose the patients that you want the students to care for and you get to watch students slowly evolve from being panicked and petrified in their first semester as they approach their first real patient to being confident and competent graduating seniors. The metamorphosis is wondrous and it is a privilege to have a part in making it happen.

4

2 UNDERSTANDING THE CLINICAL COURSE AS A CURRICULUM COMPONENT

This section provides a brief overview of the components of a nursing curriculum including mission statements, philosophy, expected outcomes, and arrangement of courses. The placement and purpose of clinical practice courses and the instructor's responsibility for implementing those courses within a schools curricular plan is the major focus.

INTRODUCTION

In many schools, clinical practice is usually only one component of a given nursing course. Other parts may include lectures, seminars, written examinations, and skills labs. In some schools, clinical practice is offered as a separate lab course and is taken concurrently with a related classroom course. In order to make the best use of the clinical time the instructor needs to understand the interrelationships between and among the clinical component, the course, and the curriculum as a whole.

In an effort to gain an understanding of the clinical course as a part of the curriculum it is probably better to try to get a grasp of the whole before examining the parts. There are several documents that can be used for this purpose. They include, in increasing order of specificity, the school's mission statement, philosophy, conceptual framework, statement of purpose, program outcome objectives, narrative description of the curriculum, curriculum plan, course descriptions, and a syllabus for the particular course you will be teaching.

PHILOSOPHY

The philosophy of a school of nursing represents a set of beliefs held by the faculty. Traditionally, the philosophy contains statements related to what the faculty believe to be the essence of the nature of the individual person, the structure and function of society, how health should be defined, the nature and role of nursing in relation to both the individual and the society, and the manner in which teaching and learning occur. The philosophy serves as the foundation for the curriculum. Terminology and definitions used in the philosophy can be expected to appear in course descriptions and to influence the sequencing and degree of emphasis given to various aspects of nursing content. For example, if the definition for health includes terms such as "adaptation" or "health-illness continuum", then these terms should appear in clinical courses and be incorporated into the nursing assessment tools students use as a part of their nursing care plans. If the philosophy indicates that professional nurses are expected to assume leadership roles in health care, then it should be possible to see how the leadership role is introduced and developed within the courses in the program.

After reading the school's philosophy take some time to think through your own beliefs about the nature of the individual and of society, your definition of health, what you see as the role of the nurse and the scope of nursing practice, and what you believe about the teaching and learning process. Unless you are a recent graduate from the school for which you will be teaching, you can expect there will be some differences between your beliefs and those contained in the school's philosophy. The extent of differences is likely to affect the degree of ease or difficulty that you may have in implementing the curriculum that is based on the school's philosophy. It is not necessary to be in full agreement with a school's philosophy in order to teach in its academic program, but it is necessary to be willing and able to incorporate the school's philosophy into your teaching strategies and to help students incorporate it into their nursing practice. It is acceptable to share your own beliefs with students as long as you explain that the opinions you are sharing are your own and do not necessarily reflect the views of the school of nursing. If you choose to take this action, be prepared to discuss the rationale for your views so that the students can understand that there is room for diversity of opinions related to nursing practice. Students should not feel pressured to adopt the instructor's point of view in order to succeed in a clinical course.

The conceptual framework, if the school uses one, is related to the philosophy statement and like that statement it can vary widely in length and complexity. A conceptual framework can be defined as a statement of the theories, concepts, or principles that serve as the organizing framework for the curriculum itself. For example, a conceptual framework might be one of the published nursing theories such as Orem's Self Care Model. If an established theory is used, the terminology, definitions, and concepts of the theory will permeate the clinical courses and will be evident in course descriptions and in the format used for nursing care plans. Another common practice is to base the organizational framework on a broad concept such as health promotion. When this approach is used the concept is usually defined operationally and its application to nursing is explained.

Terminology and definitions used in the philosophy statement should be evident in the conceptual or organizing framework. Some schools of nursing have devised graphic drawings or models to represent the relationships between the various components of the framework. These graphic representations are often accompanied by narrative descriptions that further explain how the model is used as a base for the curriculum. Some schools use a narrative statement without any graphic representation to describe their organizing frameworks. Regardless of the simplicity or complexity of the school's model or the format of its description, most organizing frameworks include familiar concepts such as the nursing process, critical thinking, evidence based practice, communication, caring, and leadership.

In addition to program outcome objectives, many schools of nursing have developed what are referred to as level objectives. These objectives describe the level of achievement expected at the end of each semester or each year of the program. They bear a direct relationship to the outcome objectives and serve to guide faculty and students in defining expectations at the beginning, middle, and end of the program. For example, when an outcome objective indicates the graduate will apply the nursing process to the care of individuals, groups and communities, a sophomore level objective might require that the student apply the nursing process to the care of individuals with clearly defined health care needs, a junior level objective might state that the student applies the nursing process to individuals and families with complex health care needs, and a senior level objective might indicate that the student applies the nursing process to individuals, groups and communities with multiple and complex health care problems.

NARRATIVE DESCRIPTION

A narrative description of the curriculum can provide an explanation of how the philosophy and conceptual framework are put into practice. It also explains the sequencing of courses and clinical activities. The most common pattern allows students to build on prior knowledge while progressing from simple and predictable to complex and unpredictable nursing situations. Sometimes the narrative description is included in the school's catalog; sometimes it is included in recruitment materials. If a narrative statement is not readily available, seek an explanation of how the curriculum is implemented from the course coordinator or from a faculty member who has taught in the program for several years.

COURSE SYLLABUS

The syllabus provides the most complete picture of the course. It includes a description of the course, the number of class and clinical hours per week, a listing of the learning objectives to be achieved by the completion of the course, a list of the topics to be included in each class meeting, the course requirements such as tests, papers, and care plans, and the learning activities such as assigned readings, clinical practice, and clinical conferences. Some faculty members include evaluation forms for use in the clinical setting as part of the course syllabus, while others distribute clinical learning objectives and clinical evaluation forms as separate items. The course syllabus provides a complete picture of the learning expectations and the distribution of the workload for both faculty and students

Read the syllabus carefully. Make a note of any questions that come to mind regarding the clinical practicum component of the course. Consider how this course contributes to the achievement of program outcome objectives. It helps put learning expectations in perspective if both faculty and students are aware of how their current efforts fit into the total program picture.

For beginning level courses, it is especially important to pay attention to the sequencing of both classroom and skills lab topics so that they can be incorporated into the clinical activities as appropriate. In advanced level courses it may be necessary to clarify how the current clinical course builds on previous courses. If the syllabus does not come with a complete set of handouts related to clinical practice, obtain these documents before the first clinical assignments are made.

The clinical instructor should analyze the relationship between the objectives for the course and the objectives for the clinical component. The objectives serve to define the level of expectation for student learning. They should be related to the overall course objectives. This relationship is obvious in clinical based courses such as Care of the Child. The relationship may be more difficult to discern in integrated courses with titles such as Nursing IV or Nursing Process 301, especially when students who attend the same classroom lectures are assigned to different types of clinical settings. The less obvious the relationship, the more important is the communication between and among the faculty teaching the course. Students are alert to discrepancies in the level of expectation among clinical instructors within a given course. While total equality of assignments is not possible in a single setting (let alone in a variety of settings), students enrolled in the same course do have the right to expect assignments that require comparable time and effort.

CONTRACTS WITH CLINICAL AGENCIES

One final document that merits attention is the written contract that exists between the school of nursing and the clinical agency. These contracts are often done in a standard format and are written in largely legal terminology. They spell out the rights and responsibilities of the school and the agency and the limits of responsibility of both the school and the health care agency for the students learning experiences and for the care of patients. They may also specify health status and immunization requirements for students and faculty; requirements for proof of licensure and liability insurance for faculty; and requirements related to CPR certification for faculty and for students. Many health care agencies now require drug screening and criminal background checks for their employees and for all students and faculty who are seeking clinical practice placements. Contracts usually contain statements related to the agency's lack of liability for expenses related to any medical care provided to the student in the event of accidental injury or sudden illness. Some schools of nursing require that students provide proof of health insurance as a condition of enrollment.

These contracts are generally reviewed and renewed annually. Short-term observational experiences in selected settings may be arranged under a simpler letter of agreement that spells out the learning objectives and provides permission for students to be present in the agency. It is risky to allow students or faculty to participate in clinical experiences in any agency or institution with which the school does not have a current, signed agreement on file.

ANALYZING THE RELATIONSHIPS

A review of the documents described above should give you a general idea of how the course you will be teaching contributes to the program as a whole. It is likely that even after careful reading of the materials related to the program and the course you will have additional questions about specific details. These questions should be referred to the course coordinator. If the clinical course is taught by a group of faculty, make every effort to attend planning meetings. The group discussions will add to your understanding of the course and of your own role as one of the clinical instructors. Before beginning your clinical teaching plans, it would be a good idea to check your perception of what you can reasonably expect from the students with a faculty member who is familiar with both the course and the curriculum as a whole.

CLINICAL PRECEPTORS

Preceptors who work with student nurses will be far more effective in their roles if they are fully aware of the information outlined in this section. In a sense, the preceptor is the clinical instructor for those hours when she or he is with the student. The greater the preceptor's understanding of the curriculum and the specific learning objectives, the greater will be the chances for a positive experience for both the student and the preceptor.

For those staff nurses who, as preceptors, are orienting new staff nurses, the focus and responsibilities are a little different. While the preceptor for the student nurse makes a short-term investment in the preparation of a future professional, the preceptor for a new staff member is making a long term investment in a professional colleague. Both the risks and the rewards are greater for those who are chosen to be teachers and role models for new staff members.

The clinical agency's philosophy and purpose statement serve as the basis for orientation programs for new staff. Agencies differ in how such programs are designed and implemented, but, as a rule, the orientation period is intended to acquaint the new staff member with the agency's philosophy, policies, procedures, and expectations. Where formal orientation programs are used, the preceptor needs to be as aware of the content and focus of the orientation program as a faculty member does of the curriculum. In the absence of a formal program for the preceptor to follow, the job description for the position and the employee evaluation form are probably the best resources to use in planning for the orientation of a new staff nurse.

3 ACQUAINTING YOURSELF WITH THE LEARNERS

This section addresses the characteristics of learners. The term learner is used to refer to students or to new staff nurses. Factors to consider about the learners as a group are presented first, followed by a discussion of factors that may influence learners as individuals.

LEARNERS AS A GROUP

There are several characteristics of the learners as a group that will influence the planning for their clinical experiences. For example, the types of nursing courses and clinical experiences the students have had prior to the course in which they are enrolled. Obviously, beginning level students will need considerably more structure and support than senior level students. The differences between second and third semester students will be less distinct regardless of the clinical site. If the course you are teaching has clinical courses among its prerequisites, then it is reasonable to expect that all students in the clinical group have met at least the minimum requirements for those previous courses. Unfortunately, a passing grade in previous clinical courses does not provide tangible information about the actual competence or confidence level of any one student or of the group as a whole. The learners themselves are the best source of this type of information.

Another factor to consider is whether all of the students in the clinical group have had the same type of prior clinical experiences. In many schools of nursing students are divided into subgroups for clinical courses; some students may be in maternity or pediatric settings, while others are in adult, psychiatric, or community health settings. As a consequence, there is considerable variety in the ability and expectations each student brings to the new clinical setting. Even in schools where all members of a class have the same type of clinical experiences in the same sequence there will be variations depending on the type of agency in which the individual students had their clinical practice and the teaching styles of their previous clinical instructors. The most efficient way to determine where the students have had their prior clinical practice experiences is simply to ask them. This can be done verbally or by having each student list previous clinical agencies and instructors.

It is important to know if this is the first time these students will be in a clinical setting like the one you will be using and if they will have additional experiences in this type of setting at a later time. Suppose that students are assigned to a maternity setting for three or four weeks early in their program of study to focus on normal maternal and newborn characteristics while also practicing communication techniques, basic assessment, and comfort and hygiene skills, and, that these same students will return to the maternity setting for three or four weeks in their senior year to focus on care of high risk mothers and newborns. The types of experiences needed by these groups of students will be different. But if the clinical instructor and the nursing staff are fully aware of the learning objectives, it should be possible to provide each group with appropriate learning experiences without overwhelming them as beginners or failing to challenge them as advanced students. Suppose, on the other hand, that students are assigned to the maternity setting for seven to ten weeks for their entire maternity clinical experience at either the junior or senior level. Planning for these students will vary from semester to semester depending upon how far along they are in the nursing program as a whole but will still need to include the full range of maternal and infant care from normal through high risk.

LEARNERS AS INDIVIDUALS

The amount and type of information the clinical instructor or preceptor obtains about the individual learner is largely a matter of personal preference. There are no norms to follow. Some faculty like to know a lot about the individual learners, while others prefer to confine their knowledge about the individual to that necessary to help meet the clinical learning objectives. Faculty have access to student records and may, if they choose, review a student's academic record and prior clinical performance evaluations. If you choose to review a student's records, remember, that like patient charts, student records must be treated as highly confidential material. Information cannot be shared without specific permission from the student. Regardless of the depth of information you choose to seek, it is important that as an instructor you be aware of your own attitudes toward individual learner characteristics. In years past student nurses were a far more homogeneous group than they are today. The proportion of "non-traditional" students is rising and this fact will require some adjustment in the assumptions made by instructors.

The age of students may vary from the late teens and early twenties to the mid-fifties and beyond. While the older students bring more life experiences, their knowledge of nursing rarely exceeds that of the younger students. The older students tend to be very goal oriented and have little tolerance for written assignments which seem repetitive or which do not have an obvious and direct correlation to their learning goals. In addition, they are more likely to raise questions about policies and procedures in both the school and the health care setting. Young and inexperienced instructors may be less comfortable with students who are well into middle age and who may remind them of their mothers or their mother's friends.

Despite increased recruitment efforts, males remain a minority within nursing. Whether your previous experience in working with male nurses has been limited or extensive, take a few moments to examine your attitudes toward men as nurses. While most faculty would deny that they harbor gender bias, many still use the term nurse to refer to female nurses and preface the word nurse with the adjective male when referring to nurses who are male.

Racial and ethnic biases are other attitudes that may have to be confronted. (What assumptions do you make when you see an African-American, a Hispanic or a Caucasian person in a uniform, a lab coat, or standard scrubs? Are you more likely to think "there goes a supervisor", "there goes a staff nurse", or "there goes a nurse aide"?) Examine carefully the ways in which you respond to students whose race or ethnicity is different from your own. Expectations related to course objectives should not be raised or lowered on the basis of race or ethnicity nor should students be assigned to patients on the basis of race or ethnicity except to assure that all students have opportunities to care for patients from a multiplicity of cultures and races.

While gender and race are obvious to even the casual observer, other personal characteristics such as gender prefernce and marital status are not. Information about factors such as marital status, responsibility for dependent children or elders, child care or elder care arrangements, and location and type of living conditions cannot be required, but if the student is willing to share such personal information it can add to the instructor's understanding of the student as a person. These types of data need not be sought directly. Students in the initial clinical conference usually share this information freely, if the instructor takes some time to allow students to get acquainted with each other and with her or him. A non-directive statement such as "tell us about yourself" will usually yield considerable information. If individual students seem reluctant to share information, then they should not be pushed to tell more than they are comfortable sharing with the others in the group. Individual students goals and expectations can be learned in a similar manner by asking such questions as "what influenced you to enter nursing?" or "what do you see as your career goal?".

Information related to students' personal lives should not alter course requirements, but may help the instructor understand the students as individuals with differing life styles, responsibilities, and aspirations. Knowing more about the individual students can sometimes pose a dilemma. For example, if you know a student lives thirty miles from the school and has been having some difficulty with childcare arrangements, do you excuse tardiness on a clinical day? Suppose there is another student in the same group who lives in a dorm on campus and who is tardy due to falling asleep after the alarm rang. Both students were late; do you treat them exactly the same? The answer is not necessarily clear and may depend upon other information you have related to each student's prior pattern of lateness. You want to maintain openness and flexibility in dealing with learners but you must understand that while flexibility is commendable, fairness is imperative.

Students appreciate any consideration that can be given to their preferences and they also respond well to sensitivity to their anxieties and fears. One way to elicit preferences and concerns is to ask students at the beginning of a clinical course if there are types of patients or situations they think they would find particularly helpful to their learning goals and if there are particular types of patients or situations they would prefer to avoid for the present. (A suggested list of questions can be found in Appendix A.) Students who had a recent experience with serious illness, injury or even death of a significant other may not be ready to cope with the care of a patient who has a similar condition. This is not to say they will never care for such a person, it simply means they need more time to deal with their personal feelings before they will be able to provide the type of supportive care the patient is likely to need. These students can be referred to counseling if additional assistance is needed.

Questions related to health problems should be addressed privately. Request that students inform you individually if they have any health problems that might become symptomatic during clinical practice. The primary purpose for having this information is to help avoid preventable incidents. If you know that a student is diabetic, you will be more likely to assess and intervene quickly if that student begins to behave in a manner not in keeping with usual patterns. If you know that a student is asthmatic, you will be more likely to take appropriate action if that student suddenly becomes short of breath. If you know that a student has a chronic back problem, you are more likely to remind that student to use good body mechanics and to get assistance when moving helpless patients. These examples should be sufficient to alert you to the importance of being aware of any significant health problems. If you know that a student is immunosuppressed or is HIV positive, you will need to determine if any modifications of assignments will be needed to minimize their risk of infection.

INDIVIDUAL ORIENTEES

The type of information a preceptor may find helpful about individual orientees is similar to that discussed in the previous section. Because the preceptor works with only one orientee at a time and because the amount of time the preceptor and the orientee spend together greatly exceeds the amount of time an instructor spends with an individual student, it is likely that the preceptor will become more familiar with the orientee as a person.

It is essential for the preceptor to know about the orientee's educational preparation and past clinical practice experience, including practice related to the current position. This information will aid in planning the extent of orientation and supervised practice likely to be needed before the orientee can assume full staff nurse responsibilities.

The amount of personal information elicited is likely to depend upon the unique personalities of the both the preceptor and the orientee. If both are outgoing persons, much information is likely to be exchanged. If both are reserved persons, less information will be shared. A general rule of thumb is to not request more information than you are willing to share about yourself. The orientee has a right to privacy in regard to her or his life outside the workplace.

Professional colleagueship is a reasonable expectation as the new staff nurse adapts to the practice environment regardless of the similarities or differences between the preceptor and the orientee. The development of personal friendships is not necessary, but development of mutual respect is vital.

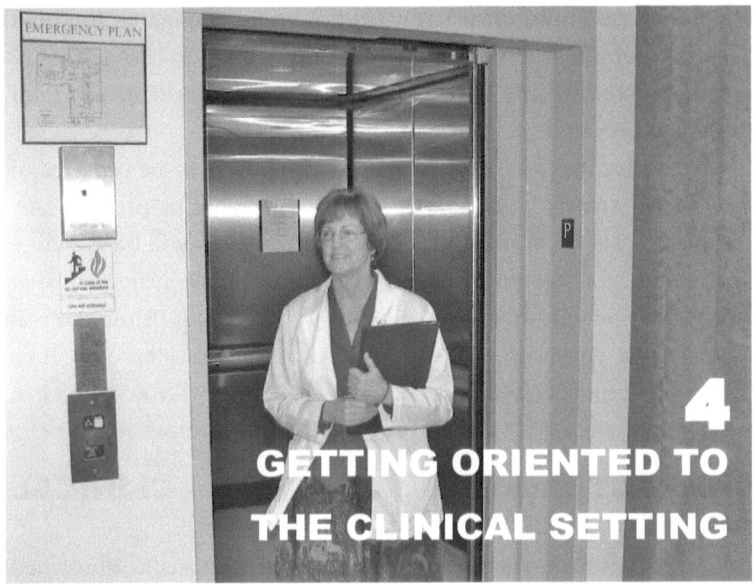

4
GETTING ORIENTED TO
THE CLINICAL SETTING

This section addresses orientation of both the instructor and the students to a new clinical setting. Topics include the general characteristics of the clinical agency, the organizational structure, the physical layout of the patient care unit, policies and procedures, and documentation of care.

Taking students into a clinical setting with which you are unfamiliar could be compared to being assigned to the neonatal intensive care unit after five years as a geriatric nurse practitioner. There are some things you could do safely immediately, but it would be some time before you or your colleagues would be willing to say you were fully competent to care for neonates. The amount and type of orientation to a new clinical unit will vary depending upon the complexity of the patient needs and your previous experience as a nursing instructor. Some people find that attending the clinical agencies orientation program for new staff is one of the best ways to become acquainted with a new agency. Others find that time spent with a head nurse or charge nurse works better for them. Still others find that spending several days or weeks in a staff nurse role is the best way for them to feel competent and confident in a new clinical setting. Some clinical agencies have specific requirements for the orientation of clinical instructors from schools of nursing. Find out what kind of orientation is required by the agency and what kind of orientation is available in the immediate area you will be using for student experiences.

Regardless of the manner in which you become oriented to the clinical area it is imperative that you be able to provide safe, competent, quality nursing care to the patients you will be assigning to students. You should be able to safely operate all equipment that is used routinely in the care of patients in the setting. Although it is unlikely that you will be providing direct care yourself, you will be making judgments about patient care when you make assignments, teach, supervise, and evaluate students in the clinical setting.

CHARACTERISTICS OF THE CLINICAL AGENCY

General information about the clinical agency can be obtained from several sources. The printed material the agency distributes to patients often contains information about the types of services the agency provides, the policies related to eligibility requirements, the non-medical services available to patients or families, and even such incidentals as the availability of parking, lodging, and eating establishments. The descriptive information included in employee orientation packets often contains references to the agency's mission, size, and services.

It is almost as important to be familiar with the philosophy and goals of the health care agency as it is to be familiar with the philosophy and goals of the school of nursing. While frankly incompatible goals that would preclude the use of an agency for student clinical experiences are rare, the instructor needs to know whether the clinical agency includes providing educational experiences for student nurses within its philosophy or purposes. Do patients who come to the health care agency understand that part of their care may be provided by students? Sometimes the consent for treatment form that the patient signs upon admission contains a statement that the patient understands that the agency serves as a teaching institution and that some of their care may be provided by students. If there is no statement related to students providing care in the agency's philosophy or mission statement the instructor should read the contractual agreement between the school and the clinical agency to determine the designated scope and limits of student participation in patient care.

The size and classification of a health care facility can be an important factor in determining whether or not it will be suitable for use as a student clinical site. The scope of experiences will be different in a small community hospital than they will be in a large urban hospital, a major teaching hospital, or a specialty facility such as a nursing home, a rehabilitation center, or a pediatric hospital. Agencies often restrict the number of students they will accept at any one time. Sometimes arrangements can be made for alternative learning experiences when numbers are restricted or when the patient census is temporarily low.

Funding for health care agencies can come from public taxes or from private sources. Agencies designated as privately funded may be further classified as either non-profit or profit. Cost containment measures and specific accounting for all expenditures have become an integral part of health care in this decade. The instructor needs to be aware of any policies that may affect the students' experiences. For example, some community health and home health care agencies are required to have care provided by the agency's paid personnel in order to bill third party payers for the services.

The location of the health care agency is important for some very practical reasons. Distance, traffic congestion, availability of parking, fees for parking, and the perceived safety of the neighborhood are important to students and faculty alike.

ROLE AND FUNCTION OF NURSES WITHIN AN AGENCY

The status of nursing within the agency is not always obvious nor is it easy to determine. Objective evidence may be obtained by determining the title and authority of the highest-ranking nurse in the organization, and noting to whom this nurse reports. Subjective evidence can be obtained by asking staff nurses about the status of nursing in the institution, how much autonomy they have in making decisions about nursing care, and how much influence they have in decisions related to policies and procedures affecting nurses and nursing care.

Information related to the organizational structure and the chain of command within the department of nursing is essential to understanding the appropriate channels of communication that are to be used by the instructor and by the students. Entering a new agency, the clinical instructor should identify the level of administrator that he or she is likely to encounter on a regular basis. What are the responsibilities of the unit administrator? If the administrator is a head nurse does she or he participate in patient care or does she or he see the role as purely administrative? Who makes the day-to-day decisions related to assignment of nurses to patients? Are there any non-nurse administrators who make decisions related to nurses or nursing care? For example, some hospitals have unit managers who are responsible for scheduling the work hours of staff nurses and/or for determining and maintaining patient care supplies.

Does the number and type of staff seem sufficient to meet the nursing needs of the patients? Are there sufficient numbers of nurses prepared at or above the educational level of the students to serve as role models for the learners? An experienced nurse of any level of preparation can assist students with selected technical skills and serve as a resource person when students have questions related to specific patients or to agency policies and procedures. However, it is not reasonable to expect a nurse prepared at the associate degree level to serve as a preceptor for a student enrolled in a baccalaureate nursing program.

REQUESTS FOR ALTERNATIVE LEARNING EXPERIENCES

Because an instructor spends a great deal of time seeking learning opportunities for students, it is important to know who is responsible for handling requests for student experiences. In some agencies all requests are channeled through a central office while in other agencies the instructor is free to call the departments directly to arrange observation experiences for students. The most commonly requested experiences outside the regularly assigned clinical area relate to observations in surgery, physical therapy, and radiology. Other experiences may include attending group-counseling sessions or participating in interdisciplinary conferences.

ORGANIZATIONAL STRUCTURE

In order to make the student experience as smooth as possible the clinical instructor should be familiar with the policies and politics of the clinical unit. What is the organizational structure, i.e., who is in charge and what is the chain of command? With whom will the instructor confer when information is needed about policies and procedures or about specific patients? Are charge nurse responsibilities assumed by the head nurse or are they delegated to staff nurses. If delegated, are the responsibilities rotated among all or only some of the staff nurses? Is the nursing staff organized around a team nursing model or a primary care model? The relationship between students and staff nurses will be different if the unit uses team nursing rather than primary nursing. Generally students become at least informal members of a nursing team and communicate primarily with the team leader. In settings that use primary nursing, students usually function as associate nurses and communicate with the patient's primary nurse. Sometimes in settings that use primary nursing, the nurses are reluctant to relinquish the care of their primary patients to students.

INFLUENCE OF PRIOR EXPERIENCE WITH STUDENTS

It is helpful to know whether or not the clinical staff has had prior experience with students. If the staff has had experience with students of the same level and from the same school then orientation of the new instructor is likely to be easier because the staff nurses know what to expect from the students. Talk to several staff members to learn their perception of the students and the school. Find out what they expect from students and from the instructor. Find out what the staff see as their role in working with students. Listen to their complaints and to their compliments.

If the unit has not had prior experience with students or has not worked with students from your school, then you will have to orient them to the role and expectations of students as they are orienting you to the clinical unit. Ideally you would like to be able to identify both the positive and the negative role models among the staff members during this session, but that will take time.

GEOGRAPHY OF THE CLINICAL UNIT

Orientation to inpatient settings should begin with the basic geography of the unit. As an experienced nurse you probably have developed some unconscious expectations about where certain supplies and equipment "should" be located. You can use your preconceived notions to your advantage while being oriented to a new unit. When supplies are located where you expected them to be you will know where to find them when you or the students need them. When supplies are kept in different locations than you anticipated you will need to devise a method for remembering the "new" location.

PATIENT ROOMS

While the patient rooms may look like a normal environment to you, try to recall how you felt the first time you entered a hospital room as a beginning level student. The amount of equipment that is standard in most hospitals today may lead the student to ponder whether a course in electronics should have been included in the list of prerequisites to nursing school. Electrically controlled beds, remote control radio and television, and the intercom to the nurses station are accepted as standard equipment in almost all hospitals, and, the near future is likely to bring the addition of computer terminals to each patient bedside. You should be able to answer the questions in the box at the conclusion of your orientation to the patient care area.

Patient Rooms

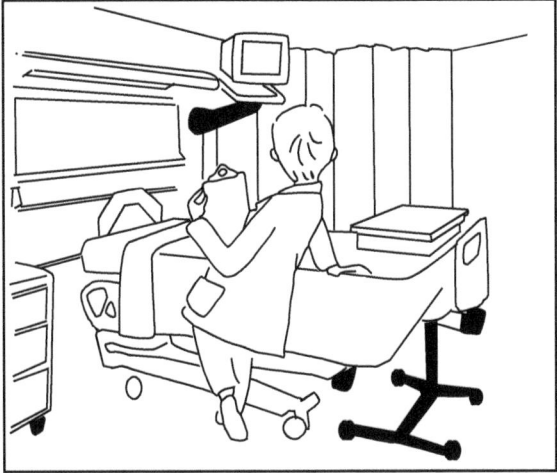

What is the numbering system and location of patient rooms?

In rooms with more than one bed how does one determine which is "Bed1", "Bed 2", etc.

Do all patient rooms have fully equipped bathrooms or do patients need to use a common tub or shower room?

How do the beds operate?

In case of electrical failure, how can the bed be raised or lowered?

How do patients "call" the nurse?

Can the nurse "call" the patient from the nurses station?

Where are the light switches?

How does the over-bed table operate?

Where is the emergency call button usually located in patient rooms?

Are patient rooms equipped with telephone, radio, and/or TV?

How does one use the phone to make an outside call?

How are heating, cooling, and ventilation controlled?

Is there an oxygen outlet and how does it operate?

Is there a suction outlet and how does it operate?

PATIENT CARE SUPPLIES

The amount and type of equipment and supplies included in the room charges for hospitalized patients varies from institution to institution. In an effort to control expenses, many hospitals have initiated greater inventory control and cost accounting mechanisms. You should become familiar with which supplies are considered part of the room cost and which are classified as separate charge items. It is important for instructors and students to behave responsibly to avoid wasting supplies, regardless of whether the items are charged or not. The box contains a list of the most common personal care supplies used in hospitals and long term care facilities.

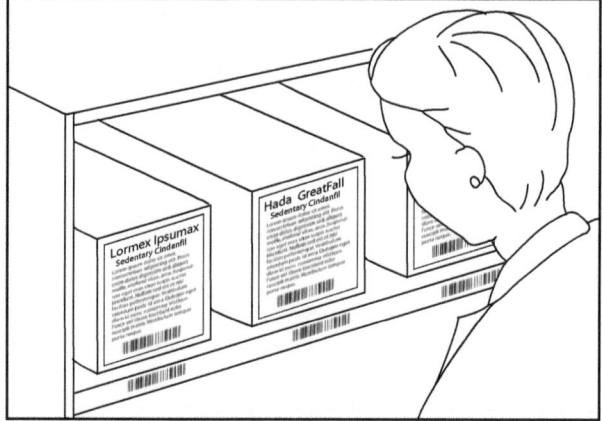

Patient Care Supplies

Where are linens and what is the composition of "linen packs" if applicable?

What is the composition of the standard bedside equipment pack and where can one find replacements such as emesis basins, soap, etc.?

Where does one obtain toothbrushes, shampoo, or lotion?

What supplies are included in the daily room charge and what supplies are charged to individual patient accounts?

What mechanism is used to account for items which are charged?

LOCATION OF SUPPLIES AND FACILITIES

To function effectively both you and the students will need to know where to find clean or sterile supplies, where to take used or contaminated equipment, and where to find creature comforts for personal and patient use. The box contains a list of the commonly asked questions you need to be able to answer.

Supplies and Facilities

Where is the "clean utility room" and what is stored there? Where is the "dirty utility room" and what is to be taken there?

Where are restrooms for staff and for visitor use?

Where is the unit kitchen and what is available from that kitchen for patients, visitors, and staff?

Where are visitor lounges/waiting areas?

What are visiting hours?

Where are the nearest public telephones and vending machines?

Where are the gift shop, the coffee shop, and the cafeteria?

EMERGENCY PROCEDURES AND SUPPLIES

Although you hope that no emergencies will occur while you are present, it is imperative that you know the location of emergency equipment and the procedure for reporting emergencies. The box contains questions related to emergencies that you should be able to answer at the conclusion of your orientation to the clinical area

Emergency Procedures and Supplies

Where is the emergency cart?
Where is the emergency phone and how does it operate?
What is the agencies code for cardiac arrest?
Where are the fire extinguishers?
What is the procedure when the fire alarm sounds?
Is there procedure for natural disasters such as a tornado?
Are there any safety precautions peculiar to the unit?

INFORMATION AND PATIENT DATA

While many clinical agencies are implementing electronic medical records it is not yet a universal practice. Some hospitals use only paper charts, some use only electronic records and some use a combination of the two. Regardless of the type of system used to record and maintain patient information, the clinical instructor and the students will need to know where and how to obtain that information. The box below contains questions you will need to be able to answer before orienting students to the clinical site.

Information and Patient Data

Where are patient charts and kardexes kept?
Where are supplies of chart forms?
If electronic patient records are used, how can students obtain access?
Where are reference books kept?
Where are the policy and procedure manuals kept?

OUTPATIENT SETTINGS

Orientation to an outpatient setting is similar in many ways to that for an inpatient setting. The box lists some additional questions you should consider if you are using an outpatient setting.

Both the instructor and the students should be able to find the supplies they will need to provide the type of patient care they will be providing. They also should be able to locate emergency equipment and all the paper documents related to patient care.

Imagine that you are assigned to care for several patients similar to the ones who will be assigned to the students and see if you can find all the items and information you would need to provide care for an entire shift. If you can, then you will probably be able to respond to most students' questions about where to find specific items.

POLICIES, PROCEDURES, AND ROUTINES

Once you understand the geography of the unit, the next task is to determine the unit routines. It is especially important to know the routine events that are likely to occur during the hours you and the students will be present. It will probably take several weeks to feel comfortable in a new clinical setting, but if you can answer the questions posed in the next several figures you will be able to assist your students with routine nursing care.

Outpatient Settings

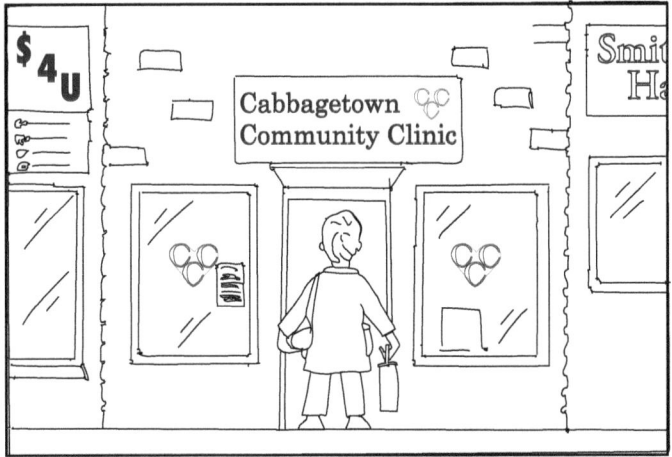

What is standard equipment in examining rooms?
Where are extra supplies kept?
Where is "dirty" reusable equipment to be put?
How are "disposables" disposed?
Where are emergency equipment and supplies located?
Who is to be called in the event of an emergency?
Where are patient and staff restrooms?
   Where are phones which may be used by patients? By family members?

CHANGE OF SHIFT REPORT

The change of shift report can vary from a joint meeting between all staff completing a shift and all staff beginning the next shift to each individual nurse listening to those parts of a tape-recorded report that describe only her or his assigned patients. A number of questions related to change of shift report are listed in the following box.

Change of Shift

What time is the change of shift report?
Where is the report held? Is the space big enough to accommodate students?
Is the change of shift report live or taped?
If report is live, is the patient Kardex used as the basis?
What information is included in the report?
Who attends the change of shift report? (nurses only or all caregivers)
Who "watches" patients during change of shift report?
Is report followed by nursing rounds?

VITAL SIGNS

Routine vital signs are so much a part of the care of the hospitalized that this topic would seem to deserve a separate section. Although there are some who are beginning to question of the efficacy of "routine vital signs," it is likely that it take some time before nursing will be able to change the present custom of taking each patient's vital signs at least twice a day as long as that patient remains in the hospital. The box contains some of the questions that arise when one is faced with the relatively simple task of obtaining routine vital signs.

Vital Signs

When are "routine" vital signs taken?
Who generally takes the routine vital signs?
Where are vital signs recorded?
What equipment is used to measure vital signs?
  Electronic thermometers?
  Mercury or aneroid sphygmomanometers?
  Are they portable, wall mounted, or automatic?
To whom should the student report significant changes in vital signs?

MEALS AND OTHER NOURISHMENTS

Fortunately, in most hospitals and long-term care facilities, meals are delivered by dietary assistants and meals are ordered by a unit clerk. Nevertheless, it is helpful to know the answers to the questions contained in the box.

Meals

When are meals delivered?
Who takes meal trays to the patients bedside?
Who removes the trays from the bedside?
Where are used trays to be put?
Who assists those patients needing help with meals?
How does one arrange for late trays or delayed meals?
How does a patient get between meal beverages or snacks?
How does one know when a patient is NPO or is on restricted fluids or a special diet?

INTAKE AND OUTPUT

Measuring and recording intake and output is often one of the first skills taught to new student nurses. The technique itself is quite simple, which may help account for the fact that recording intake is frequently overlooked. Students are more likely to be conscientious about recording intake and output if they understand why these data are being collected. They should be encouraged to examine the potential relationship between the patient's physiological disorder and the fluid intake and output data. The box contains questions that you and the students need to address.

Intake and Output

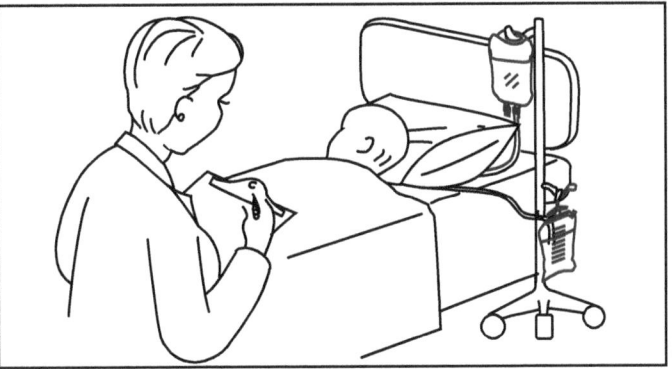

Who fills patient water pitchers?
Is water intake recorded by the glass or by the pitcher?
Who is responsible for recording intake from meal trays?
Where is intake and output recorded?
Who is responsible for determining and recording daily totals?
What type of container is used to measure fluid output?
What role does the patient and the family play in recording intake and output?

MEDICATION ADMINISTRATION

Medication administration is a common nursing responsibility that offers numerous opportunities for error. Health care agencies frequently have specific policies related to the administration of medications. The questions in the box are those that any nurse who might be administrating medications would probably ask. Specific information related to teaching and supervising students will be included elsewhere in this book.

What times are regularly scheduled medications administered?
Where are medications kept and who has access?
Are there any medications considered stock and not charged separately to the patients?
How are STAT medications obtained?
Is the pharmacy open 24 hours a day?
Who is permitted to administer medications?
Who has access to narcotics?
What are policies regarding obtaining and preparing narcotics?
If a medication error occurs, what action is taken?
How are needles and syringes disposed?
What are policies related to administration of chemotherapy?
Are there protocols for experimental drugs?

VENIPUNCTURE AND INTRAVENOUS THERAPY

Intravenous therapy is closely related to medication administration in the number of policies and procedures that have been developed. Although venipuncture is often considered a nursing function it is not universally so. You should understand clearly both the state laws and the institutional policies related to nurses doing venipuncture. Some agencies restrict the procedure to those nurses who have had additional training and supervised practice in the technique. The questions in the box relate to policies and procedures affecting venipuncture and intravenous therapy.

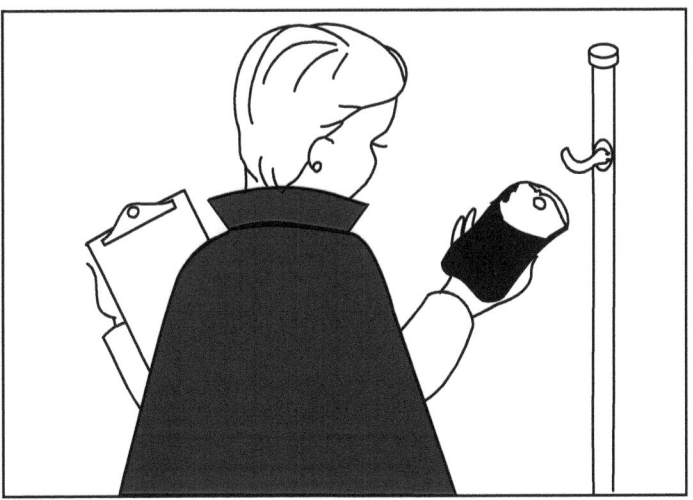

Venipuncture and Intravenous Therapy

Who may draw blood or start an IV?
Who prepares IV medications?
Who may sign for and hang blood and blood products?
What are protocols for changing IV tubing?
What are protocols for IV site care?
Who may hang TPN fluids?
What are protocols for site care for central lines?

PHYSICIAN'S ORDERS

It is unlikely that either the instructor or the students will have any direct responsibility for transcribing physician orders, but both are likely to be involved in the implementation of such orders. Therefore it is important to understand the process and procedures for obtaining and implementing physician orders. In teaching institutions it is especially important to know the agency policy

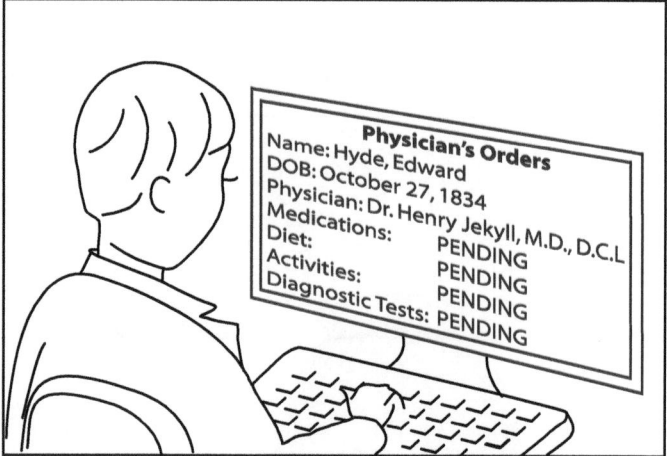

Are there any standing orders from specific physicians?

Are there approved protocols for specific diagnoses or treatments?

Who is responsible for transcribing physician orders?

Who may legally take verbal and phone orders from physicians?

What is the policy related to implementing orders written by resident physicians, medical students, nurse practitioners, or physician assistants?

related to the need for the attending physician's signature before orders written by others can be implemented. Students will need to know that the clinical unit's common practice cannot substitute legally for formal written and signed protocols and physician orders. The box contains questions related to this topic.

ROLES OF HEALTH CARE TEAM MEMBERS

Related to the agency routines is a body of information that describes the role of the nursing personnel including supervisory or administrative nurses, staff nurses, charge nurses, primary nurses, team leaders, team members, clinical specialists, and nursing assistants. Understanding their relationships will allow you to channel questions to the appropriate persons. The unit clerk or ward secretary is often an excellent source of information about the location of supplies and equipment, unit routines, procedures for requesting services or supplies, the identity of physicians and other health care team members, and the whereabouts of patients who are not in their rooms.

SUPPORT SERVICES

Size, complexity, and budgets can affect the type and level of support services available within a clinical agency. Support services include pharmacy, physical therapy, dietetics, radiology, chaplaincy, social service, laundry, central supply, and transportation. The presence or absence of these support services will determine such activities as how patients get to and from X-ray or other diagnostic studies, and how specimens are sent to the lab.

CHARTS, KARDEXES AND OTHER SOURCES OF PATIENT DATA

One last category of information about the clinical unit that is important to the instructor and to the students is that of patient information. You and the students need to know where patient charts are kept and what they contain. Some hospitals use daily flow sheets at the patient bedside for recording vital signs and other observations and treatments. In some institutions all health care personnel record their comments on the same patient progress notes while in others physicians, nurses, nutritionists, and physical therapists all use separate forms for charting. Nurses notes may be written in one of several formats. The most popular form is that of the SOAP notes, but some hospitals still use a narrative format and some use checklists with narrative options. The advent of electronic patient records is making significant differences in how information is added, accessed, and secured.

Are there Kardexes and how are they used? Frequently they are treated as short-term records and used only as a quick reference for planning care. Ideally, the Kardex should be kept sufficiently up to date that a float nurse could provide effective care based solely on the information contained in the Kardex, although they are usually not considered part of the patient's permanent record.

Where are nursing care plans or care maps kept? Are they part of the permanent record or part of the temporary record? Who initiates care plans or care maps? Do all personnel who provide nursing care have easy access to the nursing care plan? May students initiate care plans and contribute to existing care plans?

Who initiates referrals? What agencies are used for follow up after patient discharge? Do either narcotics signout sheets or nurses notes need to be cosigned by an instructor if signed by a student? Are there any restrictions related to student learning experiences?

OTHER GENERAL INFORMATION FOR STUDENTS

The orientation of students to the unit should be done in cooperation with the nursing staff. Students will need the same general information about the clinical unit as the instructor. In addition, students will need orientation to course expectations as they relate to clinical practice. It is important to inform students about your expectations for them when they are unable to come to an assigned clinical time. It is helpful if the students are given both the agency's and the instructor's phone numbers. If possible the instructor should obtain the home and/or cell phone numbers for each student. Having phone numbers allows both students and the instructor to contact each other in cases of sudden illness or other emergencies that would interfere with either party's attendance.

If the school has a uniform dress code, the instructor needs to be aware of the code and of the procedure for dealing with violations. If the clinical agency has policies related to CPR certification, liability insurance, immunizations, or health status, the instructor needs to know whether compliance is monitored by the school or by the individual clinical instructors. If ID badges are required for students to obtain parking and/or access to the clinical units, the instructor is usually responsible for obtaining and collecting such ID materials.

COMMUNITY HEALTH AGENCIES

When the clinical site is a community health agency where students are focusing on the role of the community health nurse, there are a number of variations to the general orientation suggestions given above. The new instructor should become familiar with the agency policies, politics, and protocols. The geography of the community health setting extends far beyond the walls of the health center, so it is important for the instructor to know the geographic boundaries of the district. A tour of the area to be used for student experiences will help in decision making related to home visits and to community assessment assignments. A good map is essential. The map can be the traditional paper map or it can be one obtained through one of the web-based mapping services. If students have cell phones, they should be instructed to carry them with them during community health clinical. If the agency has any particular personal safety programs or policies, both the instructor and the students should be well informed of the particulars.

The instructor should make every effort to "know the neighborhood" which the health center serves. Obviously the instructor cannot supervise each student experience in the community in the same manner that it is done in an in-patient setting, but she or he should establish a mental picture of where the students are and what they are doing. Supervision of students in clinic setting, on the other hand, can be quite similar to that in hospitals in terms of the amount and type of direct supervision and teaching.

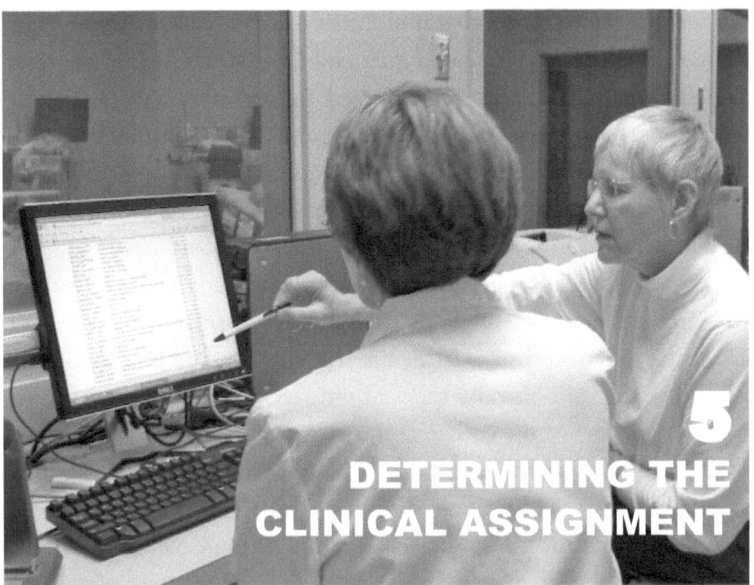

5. DETERMINING THE CLINICAL ASSIGNMENT

This section explains the process that a clinical instructor might use to choose patients for assignment to students. The setting is assumed to be an in-patient unit in either an acute or long-term care facility providing services to physically or mentally ill adults or children. A second assumption is that the instructor is choosing the patients for assignment at least the day before the scheduled clinical practice time. In settings such as obstetrical units or day surgery the instructor often chooses the patients for assignment only minutes before the students arrive. In such circumstances the steps outlined in this section are used within a shorter time frame. Although there would be additional variations in out-patient facilities, many of the guidelines are applicable to any setting in which an instructor is seeking learning experiences for student nurses.

IDENTIFYING POTENTIAL CANDIDATES

Using the clinical learning objectives as a frame of reference the first step in choosing patients for assignment to students is scanning the roster of patients to see which ones might be appropriate. If the unit uses a Kardex format, it is likely to be the most efficient source for initial patient information. Generally the

Kardex will contain demographic information such as name, age, address, occupation, marital status, and date of admission; physical data such as height and weight, need for assistive devices such as hearing aids, eyeglasses, or crutches; health status information such as general physical condition, restrictions related to diet, fluids, or activity, the medical diagnoses, and surgical plans or findings; and current medical, surgical, and nursing orders.

At this stage of the selection process it is advisable to identify as many potentially assignable patients as possible. Consider not only the types of technical skills the students need to practice in the clinical setting but also the types of written work the students must complete as part of the course requirements.

Beginning level students will fare better if they are assigned to patients whose physical conditions are relatively stable and whose mental and emotional status will permit normal communication. Advanced level students need opportunities to apply their knowledge in complex situations that allow them to use critical thinking and problem solving techniques.

If the learning objectives for beginning level students include completion of a written interpersonal process recording (IPR), then the patients selected should be able to interact verbally and rationally. If on the other hand the student has mastered basic communication skill, then learning may be enhanced by assigning a patient with a speech or hearing problem or someone whose thinking or interaction patterns are disturbed. Unless discharge teaching is a major objective, try to avoid choosing patients who are likely to be discharged within the next 24 hours. It is frustrating for students to prepare for an assignment only to find that the patient is going home in two hours, or worse yet, was discharged the evening before. Unfortunately in today's managed care environment a patient's stay may be shorter than anticipated.

Although there is great merit in expecting students to provide "total patient care" within the limits of their knowledge and experience, there will be times when only part of the learning objectives can be met if one adheres slavishly to the total patient care axiom. There may be times when it is advantageous to permit a student to provide only selected aspects of care. For example, a student could change a complex sterile dressing on a patient who is assigned to a staff nurse, or a student might be assigned to administer medications to several patients in order to gain skill and confidence in performing this procedure.

CONSULTATION WITH NURSING STAFF

The second step in the assignment process is to consult with the nursing staff. This means talking to the head nurse, the charge nurse, the team leader, and/or the primary nurses. Those nurses who have been caring for the patients are your best source of information and advice regarding the appropriateness of assigning a student to a particular patient. Remember that your responsibility is to find the type of patient most likely to meet the students' learning objectives. Not every patient who the staff describes as "interesting" or "challenging" may be appropriate for students. On the other hand do not automatically dismiss from consideration those patients who the staff describe as "difficult" or "demanding". Sometimes such patients benefit from the individual attention that a student can provide, while the staff benefits both from the temporary break and from the different perspective the student may provide. The decision to accept or reject a suggested patient rests with the instructor, but an open discussion of the rationale for acceptance or rejection is important to the future of the instructor-staff relationship.

If you describe the types of learning experiences you are seeking, it will assist the staff in making better suggestions. Unless the staff person is one you have worked with before, it will probably be necessary to describe the level of student and the general learning objectives. On those occasions, when it appears there are no appropriate patients, the staff may be able to identify secondary health problems or patient needs that can be related to the learning objectives. For example, if you are using a general surgery unit and the focus for the class that week is interferences in locomotion, the staff are likely to know which patients have underlying conditions such as arthritis, amputations, chronic back pain, or residual muscle weakness from a previous stroke. Ideally, your consultation with the staff will yield a list of patients that exceeds the number of students for whom you have to make assignments. Keep in mind that you should be able to care for any of the patients on the list and none should be beyond the capabilities of students.

MATCHING PATIENTS AND STUDENTS

The third step in the assignment process is to match the patients to the students. The first time you plan for a group a random assignment is fine, but after that you should be aware of the experiences each student has had and those that he or she still needs to meet course and clinical objectives. Some instructors keep their weekly assignment sheets in a notebook (or a PDA) that they take to the clinical unit when making assignments. Others prefer to keep a separate roster of student assignments either as a paper copy or in a computer data base or spreadsheet. Pertinent data include the patient's age, gender, medical diagnosis, nursing diagnosis, medications, treatments, activity level, and any significant psychosocial information. The method of keeping the data is not important. What is important is to have information about prior assignments available so you can ensure that each student has a variety of patient experiences and a reasonable opportunity to meet the learning objectives of the course

A second consideration is matching the patient's care needs with the student's learning needs and level of ability. The cumulative roster of previous assignments, discussed above, becomes a valuable resource for this task. At this point in the selection process the list of potential patients should include only those who are appropriate for assignment. The goal is for students to be challenged, but not overwhelmed, by their clinical assignments. Ideally, each student will be able to build on previous learning while new gaining insights and developing psychomotor skills.

In long-term care settings one should try to maintain a balance between assigning a student to one or more patients for the duration of the clinical course and providing the student with short-term experiences with other patients in the setting. This approach can have the effect of allowing the student to gain the depth that can come from a long-term patient relationship while still allowing an appropriate breadth of experiences. This approach also enhances the students understanding and appreciation of the role of the nurse in long-term care settings.

Some review of each patient's chart, whether paper or electronic, will be necessary in order for you to make informed decisions. The amount of time you spend with charts will depend upon how much information you think you need and the rapidity with which you can glean information. Gather as much data as you would collect if you were to care for the patients yourself. The following types of information about each assigned patient will prove useful in responding to student questions: age, gender, education, occupation, marital status, available support systems, living conditions (i.e., house, apartment, urban, rural), size, (i.e., height and weight), date of admission, admission diagnosis, general condition, diagnostic tests completed and planned, surgery planned or completed, concurrent or chronic physical or mental disorders, current medication and treatment orders, diet, fluid, vital signs and treatment schedule, response to hospitalization, nursing diagnoses, projected date of discharge, if known, and names of physician, primary nurse, and social worker.

Some instructors prefer to see the patients themselves before deciding to include them among the potential assignees. Although this practice is clearly a useful means of gathering data, it can be time consuming and one needs to consider the cost-benefit ratio of this approach to patient selection. The one instance where it is advantageous, regardless of the time involved, is when you are choosing patients for the students' very first "real" clinical assignment. It can be devastating for a beginning level student to enter a patient's room for the first time only to be told in no uncertain terms by the patient that they don't want to be cared for by a student. After their initial experiences most students are able to understand and accept that patients do have the right to refuse care from a student and that the patient's preference is not a reflection of the student's ability. Regardless of the level of student, when the staff nurses have any doubts about whether particular patients might be willing to have care provided by students, it would be advisable to confer with the patients directly about their preferences.

INFORMING STAFF OF STUDENT ASSIGNMENTS

The fourth step in the assignment process involves informing the staff about which patients you have chosen for the students. This includes clarifying exactly which aspects of care the students will be doing and which aspects they will not be prepared to provide for the assigned patients. For example, if the students are beginning level they probably would not be giving medications or doing sterile dressings. The staff needs to know this so that they can arrange to provide whatever care the students cannot provide. When staff are informed that students will be doing procedures for the first time they often try to provide additional support and encouragement. In some instances staff may provide assistance and supervision for students who need additional practice in performing selected psychomotor tasks. If the staff is to provide such supervision, it must be arranged in advance and must be allowable within the contract the school has with the clinical agency.

As a matter of courtesy to the staff you should try to distribute the student assignments evenly among the teams, primary nurses, etc. This has the effect of distributing the workload more equitably. It is important to remember however, that the quality of the students learning experiences holds a higher priority than the even distribution of staff-student contacts.

FINALIZING THE ASSIGNMENT

The last factor to consider before finalizing the student assignments is that of the composite effect. If all students are assigned to patients with complex nursing care needs, you may not be able to provide the teaching, consultation, and, supervision needed by each student. On the other hand, if all of the assignments are well within the limits of the students' prior experience and abilities, then the opportunities for both teaching and learning are limited. The same guidelines apply to the composite assignment as apply to the individual student assignment--it should be challenging without being overwhelming.

POSTING THE ASSIGNMENT

It is advisable to use a standard form for posting assignments. The use of a form promotes consistency in the types of information available to staff and to students, and, also promotes recognition of the presence of students in the clinical setting. In order to protect the patient's right to privacy avoid listing the diagnosis with the name of the patient, and post the assignment in a designated area that is not visible to visitors. A sample form is included in Appendix A. The form includes the instructor's name and phone number so that she or he can be contacted if the assignment needs to be changed. The date and time the students will be present is included so that staff know when they can expect students to be with patients. If the students are doing less than a full shift it is better to list only the time the students are actually available for patient care experiences and to omit, or list separately, the times used for any pre or post conferences. The clinical learning objectives are listed as well as the specific nursing care tasks that all students in the group can be expected to perform. The patients' names and room numbers are listed beside the students' names, and a space for "special data" which can include additional information peculiar to the specific patient. Examples of such special data include phrases such as "will go to PT with patient" or "will do all care except central line dressing".

Listing the patients in a logical sequence according to the location of their rooms serves two purposes. First, it makes it easier for the instructor to organize her rounds during the clinical hours; and, second, it makes it easier for the staff to find out which patients are assigned to students.

Most nursing units have specific places where assignments are posted for staff. Posting the student assignments near the staff assignments will help all staff be aware of which patients will be cared for by students and should help clarify what the students will or will not be doing for each patient.

A sample form for posting student assignments in the school of nursing is included in Appendix B. Note that this form omits the patient name but includes the diagnosis and other information that the student will need in order to prepare for the clinical lab. This form is designed for use with beginning level students who do not yet do pre-assessment visits to the patients the evening before the actual clinical day. For more advanced students who are expected to assess the patient and to prepare care plans, it is probably not necessary to post assignments in the school of nursing. Since these students will need to go to the clinical agency in order to begin their preparation, they can determine their assignment by reading the form posted at the nursing unit. Some faculty notify students by email or through password protected, web-based bulletin boards. Electronic posting, like paper posting, should never include personal information that could reveal the identity of patients.

INSTRUCTOR PREPARATION

The instructor will need a mechanism to maintain awareness of the patient care load assigned to the student group as a whole. A sample form for instructor use is included in Appendix C. This form contains patient information and must be handled as confidential information. Its purpose is to give the instructor a quick reference and to provide a form for recording pertinent observations of student performance. If the patient listing on this form parallels the form posted on the clinical unit, then patients will be listed in a geographically logical manner thus saving many unnecessary steps on the innumerable rounds the instructor will make in the course of the clinical day.

The final step in making clinical assignments is really the instructor's preparation for the clinical lab. The instructor should review any diagnoses, procedures, medications, or treatments with which she or he is unfamiliar. For the beginning instructor the author offers the same advice she gives to beginning students. Get a good night's sleep and eat breakfast before going to the clinical area.

COMMUNITY HEALTH SETTINGS

Determining the clinical assignment for students in community health settings follows the same general guidelines listed above. The clinical instructor should consult with the nurses who have been responsible for the families that the students will follow during their course of study. Time and distance should be considerations for both students and faculty. Time spent in cars contributes little to learning. Also, great distances between students reduces the amount and type of supervision the faculty can provide. The argument that "travel is a part of real community health" can be countered with the reminder that students are learners who are paying for the clinical experience and who have much to learn in little time. Learning objectives need to be reviewed carefully to determine what kind of learning activities are needed to meet those objectives. The instructor should strive to achieve a balance between clinic and home visit experiences and to consider the place of experiences in occupational settings, schools, and correctional facilities in the overall learning experiences of students studying community health.

PRECEPTORS

The general guidelines outlined in this chapter can also be used by preceptors who are working with student nurses. Preceptors have the advantage of already having information about the patients. If they have an understanding of the clinical learning objectives and are acquainted with their students, then making the patient assignments should not be a problem.

Preceptors who are orienting new staff nurses can adapt the guidelines in this chapter to fit their particular clinical settings. The general principle of beginning with the simple and moving toward the complex is particularly useful in planning clinical assignments for the new orientee. Generally speaking, the goal is to move the new staff person to full functioning as soon as possible. The rapidity with which the new staff person can move toward a full patient load will depend upon prior experience, level of confidence, and the complexity of the clinical setting. Many clinical agencies combine general orientation to the institution with orientation to the specific nursing unit. In these circumstances the preceptor needs to be aware of what the orientee is learning in "class" so that the clinical orientation can parallel the general orientation.

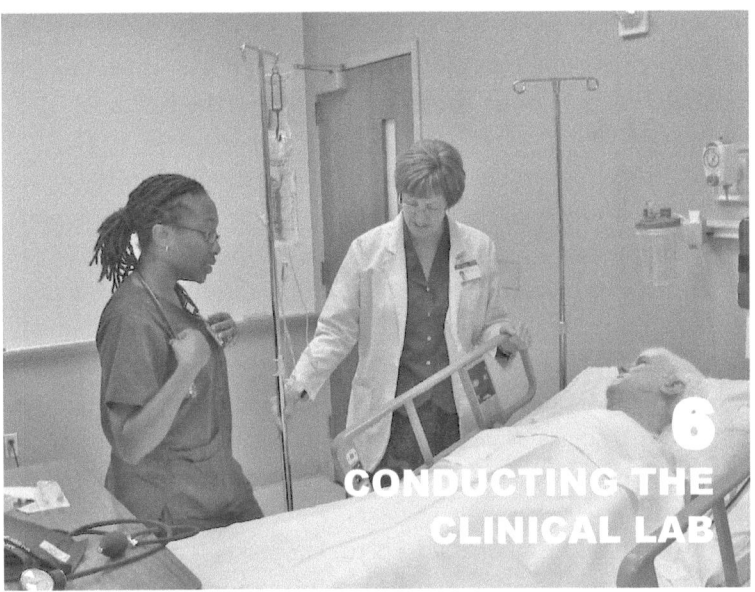

CONDUCTING THE CLINICAL LAB

This section focuses on the role of the clinical instructor during the period when the instructor and the students are in the clinical setting. It includes guidelines for pre-clinical conferences, organizing teaching and supervising activities, and conducting the post-clinical conference.

PRECLINICAL DATA COLLECTION

In order to avoid unpleasant surprises it is a good idea for the instructor to get to the clinical area about 15 to 30 minutes before the students are scheduled to arrive. This allows time to scan the patient roster to be sure all of the assigned patients are still on the unit. A quick look at care plans and or doctor's orders should identify any changes that have occurred since the charts were reviewed the previous day. A brief conference with those staff nurses who are responsible for the patients assigned to the students should serve to complete the instructor's data collection.

Some people prefer to get their updating by attending the change of shift report. This is a preferable alternative if it is workable. The use of audio-taped reports, team reports in which two or more nursing teams are receiving report simultaneously, or the practice

53

of giving report to primary nurses all tend to make the change of shift report less useful to the nursing instructor who is legitimately interested in only those patients whom she or he has assigned to students.

The role of the instructor in gathering information for students will vary with the level of student. It is probably not useful for beginning level students who often have limited clinical hours to spend precious minutes listening to change of shift report. In most cases they will not understand the terminology being used and will not be responsible for any of the care given to patients other than the one(s) assigned to them. After the initial one or two clinical courses, when students have some grasp of terminology and the scope of the role of the professional nurse, they could be expected to participate in choosing their own patients for assignment, collecting their own data, and attending all or at least pertinent portions of the change of shift reports.

PRECLINICAL CONFERENCE

The pre-clinical conference provides a mechanism for both the instructor and the students to establish the learning focus for the day. Whether it is done as a group activity or one-on-one with the instructor, the pre-conference should be brief. It is a time for giving and getting information—not a time for philosophical discussions. The instructor should determine whether students understand what they are expected to do with and for their patients. If students were to bring care plans with them these plans should be reviewed briefly. Learning objectives should be identified for the group as a whole and for the individual students. The instructor can use her own student and patient data sheet to identify which students are likely to need assistance or supervision with new procedures. Tentative plans can be made for sequencing events so that the instructor can be available to each student needing assistance with particular procedures. Anticipated problems can be identified and potential solutions can be offered. The point of pre-conference is to set the tone for the experience. It is important to not get bogged down in so many details that students are unable to get to their patients. If students are going to take morning vital signs it is a good idea for them to get them done before breakfast. It simply makes life easier for the patient, the staff nurse, the physician, the ward clerk and the instructor, any one of whom is likely to ask about the patients' vital signs at precisely 8 o'clock.

INITIAL ASSESSMENT OF STUDENT-PATIENT SITUATION

It is not possible to be with every student while she or he provides care to patients. At the same time the instructor bears responsibility for the actions of each student under his or her supervision. But how can an instructor promote learning and maintain patient safety when they are responsible for six to eight students, some of whom may be assigned to two or three patients? The answer lies in the application of the nursing process to each student-patient situation and to the group as a whole. You briefly assess each student-patient situation in much in the same manner that you would assess each patient as a staff nurse responsible for a group of patients. The purpose of the assessment in each case is to establish the priority of need for your attention and intervention.

The initial assessment of the student-patient situation is best made as soon as the students enter the patient's room. It is advisable to see the most complicated patients first since these students are most likely to need advice and/or assistance. If you have not met the patient previously, introduce yourself or allow time for the student to introduce you. Your role should be explained in a way that does not detract from the role and abilities of the student. Be alert to the patient's general appearance and behavior. Note the presence and functional status of any intravenous lines, feeding tubes, catheters, monitors, or other special apparatus. If the patient is in the early postoperative period check the wound site or the dressing if the student has not already done so. Observe what the student is doing and find out if she or he has any questions or problems that need your attention immediately. Inquire about the patient's vital signs and the student's plans for the next half-hour. Although these may seem like very ordinary questions the answers will allow you to determine whether the student has begun to implement the care plan, whether the student feels able to cope with the situation, and whether the patient has any abnormalities in vital signs that may need to be referred to the nursing or medical staff.

ORGANIZING TASKS AND SETTING PRIORITIES

After completing the initial assessment for each student-patient pair take a few moments to review your patient data sheet and any notes you took during pre-conference about students needing assistance with new procedures. Make a tentative plan of action giving priority to those tasks that must be done by a specified time. The most obvious example is that of medication administration and blood glucose monitoring. The beginning instructor often underestimates greatly the amount of time needed to supervise one beginning level student in the process of giving medications to one patient. It is wise to permit no more than two to four students to give medications on any one clinical day if the students are beginners. In an effort to minimize the chances for error when new students begin giving medications be sure the staff nurses know which patients will be receiving medications from students. It is helpful to mark the Kardex or medication administration record with a 3x5 card which lists the student's name, the patient's name and room number and the times the student will be giving medications. The amount of information the student should have about the medications to be given may be defined by either course or clinical objectives. If not, then the clinical instructor assumes the responsibility for determining the type of information the students should have. As a general rule the student should have the same type of information about a new medication as a staff nurse needs i.e., usual dose, route of administration, uses, side effects, toxicity, synergisms, incompatibilities, and precautions. This information need not be committed to memory but if reference materials are needed, they should be with the student on the clinical unit.

Aside from medications the next most common procedures needing instructor supervision are those that require the application of the principles of surgical asepsis and those procedures of an intrusive nature that carry any risk of harm to the patient. The instructor should supervise any procedure that poses a risk to the patient until such time that both the student and the instructor are satisfied that the student has the competence to perform the procedure alone.

It is easy to become involved with one or two students who are caring for patients with complex needs or who are performing new procedures for the first time. However, it is important to be aware of the needs and the progress of the other students in the clinical group. A very brief visit to each student in between the supervision of procedures will help keep you up to date on everyone's progress and allow all of the students to seek advice if they need it. Beginning students tend to lose track of time and frequently need to be reminded that they need to complete the care tasks by a predetermined hour. Advanced students are sometimes acutely aware of the time but need help in organizing tasks and in setting priorities. In making periodic rounds be alert to what students are doing and what they are saying to patients and to staff. Offer positive comments to those students who are progressing nicely, particularly if they are showing improvement in their performance. Correct errors of students as soon as possible but do not criticize or correct a student in the presence of the patient, the staff or other students. Except in truly life threatening circumstances it is always possible to find a private place to discuss a student's errors. Students will gain more from the experience if allowed an opportunity to point out mistakes on their own and to propose ways in which to correct the errors or to avoid such mistakes in the future. Students do not deliberately perform badly; often the mistakes are made out of ignorance or as a result of anxiety. The instructor's goal is to help the student obtain the necessary knowledge without adding to the student's anxiety.

CORRECTING STUDENT ERRORS

Students will feel more secure if they know in advance that they will not be criticized in front of the patient and others. The instructor should include an assurance during the initial orientation of students to the clinical unit that corrections will be made privately. The author has used a verbal signal approach successfully with both beginning and advanced students. The signals are simple and consist of the use of question or declarative statements. I explain them to students as follows:

If I ask a question such as, "Would you like some help making that bed?" then the choice is yours. If you would like help say so and I will help you but, if you prefer to complete the task alone say so and I will leave you alone to do it. On the other hand if I make a strong declarative statement such as, "let me get you another catheter" or "let me help you get Mr. M. out of bed," what I am saying to you is "you have contaminated that catheter and it cannot be used" or "the technique you are using to get Mr. M out of bed appears to be unsafe or is likely to cause him undue discomfort." If you misinterpret my intention I am likely to say, "I really think it would be better to get another catheter or use my help for now." Once you understand the message let me assist you and we will discuss the error as soon as the procedure is done and the patient can be left alone safely.

It has been the author's experience that students appreciate the use of an approach that does not embarrass them or diminish the patient's confidence in them.

MAINTAINING CURRENT PATIENT INFORMATION

If the clinical time occurs during a period of the day when physicians are likely to be writing new orders, it is important to be aware of any changes in the orders for patients assigned to students. For beginning level students the instructor can periodically check either the charts or ask the appropriate staff person about changes in orders which should take effect while the students are caring for the patients. Advanced students should be instructed to do this checking for themselves and to inform the instructor of changes they will be incorporating into the care they are giving to patients.

BEING AVAILABLE TO STUDENTS

It is important for the instructor to be available to the students in the clinical area, especially with beginning level students. All students should know where the instructor is likely to be on the unit, if they should need her. They can use the posted assignment sheet to find which patient rooms are likely places to look if the instructor is not present in either the nurses station or the conference room. If it is necessary for the instructor to leave the unit, then students should know when she is leaving, where she can be reached, and when she plans to return. The charge nurse and the unit clerk should have the same information. It is inadvisable to leave beginning level students for longer than fifteen minutes. If you are teaching advanced students, there is less likely to be panic if you are out of sight when the student begins to look for you, but even advanced students need to know when you plan to be off the unit. If you are responsible for students on more than one unit at the same time, set up a tentative schedule with each group of students and give them the telephone extension number of the other unit(s) where you can be found.

RELATING THEORY TO PRACTICE

If there is a universal learning objective for clinical lab it has to do with the application of theory and didactic content to the practice of nursing. One of the most efficient ways to determine if students are applying their nursing knowledge in the care of patients is the ask them to explain the rationale for the actions they are taking. The objective of questioning is to determine what the student <u>does</u> know and not to determine how much the student <u>does not</u> know. Often, students know much more than they realize when a question is first posed. If a student gives a very shallow answer, try restating the question. If a student indicates that she or he does not know the answer to a question that does not mean the instructor should immediately supply the answer and mark the student down for being ignorant or unprepared. Instead, there are several approaches that may be used. If the answer is unidimensional, such as the usual dose of a drug, tell the student to look up the answer in a reference book. If the answer is multidimensional and includes scientific principles and/or nursing theory then assume that the student knows at least part of the answer and break the question down to simpler parts that might "lead" the student to the answer. This approach helps build confidence and helps students develop analytical thinking.

It is sometimes difficult for clinical instructors and for clinical preceptors to control the impulse to teach the student everything they have learned during all of their years of schooling and practice. Any course, even a clinical course, is only one part of the total educational process for the student. It is usually safe to assume that the faculty members responsible for the curriculum have planned carefully the placement of content and clinical practice activities in order to assure that each student will be able to meet the outcome objectives before graduation.

COMPLETING AND DOCUMENTING PATIENT CARE

As the time for the conclusion of the clinical day approaches the instructor makes rounds one last time to confirm with each student that all the nursing care the student was expected to do has been completed and reported or recorded appropriately. Read nurse's notes and/or patient progress notes written by the students to ensure that they are inclusive and that they conform to the agency format. This is particularly important with a new group of students. Students who are just learning how to write nurse's notes should be instructed to write their notes on scrap paper and have them approved by the instructor or the staff nurse responsible for the patient before putting the note on the patient's permanent record. Check the narcotic record if any of the students have given narcotics, and review charting of medications, treatments, intake and output, and other nursing actions. Each student should give a verbal report of the patients' current condition and a summary of the care given during the time the student was with the patients. As soon as students are able to do so, they should be directed to collaborate with the staff nurses in contributing to written care plans for patients.

POSTCLINICAL CONFERENCE

The clinical practice period is usually followed by a postclinical conference. This conference is often considered a part of the clinical practice itself. It is usually held in a private area such as a conference room in which the door can be closed to assure privacy for the students and to assure that any patient information can be discussed confidentially. The post-conference should not be held in public area such as a cafeteria or a visitors lounge because of the risk of inadvertently violating the patients or the students right to privacy of information.

The post-conference time (30-60 minutes) is used to review the events of the day and to discuss the student's successes and problems. Care plans can be evaluated by individual students and by the group as a whole and modifications can be proposed. Students can relate the events of the day to the learning objectives that were identified in the pre-conference. The post-conference is an ideal setting to discuss the application of theory and didactic content to real patient situations. Students have an opportunity to ask questions about their patients, about the care they gave, or about the care they observed others give. Sometimes a general question such as "what did you learn today?" or "what would you do differently if you could do today over?" will get discussion going when students seem to have nothing more to say.

Post-conference can provide an excellent medium for students to learn group problem solving skills and to learn vicariously from their peers. Post-conference is not a time for the instructor to lecture or otherwise present course material that is not directly related to the days clinical experiences. The role of the instructor is to maximize the learning for each student by encouraging all students to share their own experiences in applying theory to the patient care situation.

There are times when the post-conference serves as a debriefing session particularly if one or more of the students has had to cope with a difficult or an anxiety producing situation. Allow sufficient time for the students to express their feelings and to receive support from the group members before moving the group to discussions of the preset learning focus for the day.

Try to begin and end post-conference on time. Both students and faculty members have other obligations to attend to at the end of the clinical period. If one student seems to need additional time it would be better to see that student individually than to keep the whole group overtime.

CLINICAL JOURNALS

The use of clinical journals can help faculty teach and evaluate critical thinking and clinical decision making. Clinical journals can be used to help students describe and analyze their own thinking patterns as they apply theory and concepts to actual clinical situations. In addition, the narrative descriptions of assessments, problem identification, rationale for decisions and actions, and, evaluation of actions taken can provide evidence of cognitive, affective, and psychomotor learning and of the student's insight into their own responses to the nursing role.

Journals are generally due each week. They can be either paper or electronic but they must not contain patient names. Regardless of format the clinical faculty needs to provide positive feedback as well as suggestions for improvement.

COMMUNITY HEALTH SETTINGS

In community health settings many of the same approaches discussed above still apply. Pre and post clinical conferences are used for the same reasons as in the hospital settings. Organizing each day so that the students know when to expect you to be in the clinic or when you will make a home visit with them is essential. Students need to know where you expect to be throughout the clinical day so that they can reach you by phone if necessary. By the same token you need to know where the students expect to be and when they expect to return to the center. If students are being paired up with a staff nurse it is important that the nurse know what the student is expected to do and what her role is in providing assistance and/or supervision.

7 EVALUATING CLINICAL PERFORMANCE

This section addresses the final phase of clinical teaching which is evaluation of the students' progress toward and achievement of the performance objectives. The clinical instructor's responsibilities do not end with the post-clinical conference. They include, but are not limited to, follow-up activities that are part of the clinical learning experience, Examples include evaluation of any written work which students do related to their clinical practice, maintenance of anecdotal notes and other records of student performance, periodic conferences with students for the purpose of discussing progress toward meeting course objectives, and written evaluation of each individual student's performance.

WRITTEN NURSING CARE PLANS

Typically students are expected to complete at least brief written nursing care plans for each new patient. The format for care plans can vary widely from school to school and even from course to course within the same school. Most daily care plans use some variation of the nursing process for the general framework. Whatever format is used, there should be agreement among the faculty teaching in a given course as to what constitutes a satisfactory level of performance on a written care plan. If care plans or case studies are to be given a letter or numeric grade then guidelines should be developed for both student and faculty use. Students will be able to improve their use of the nursing process more rapidly if they are given feedback that is specific enough to be helpful in correcting their inaccuracies.

Evaluating care plans and case studies is a time consuming activity. The time you invest can be justified only if students benefit from specific written feedback that serves to reinforce their correct application of theory and principles. The feedback should provide constructive criticism when they have made erroneous judgments and offer suggestions for improvement when they have done less than was expected.

Care plans and case studies can provide students an opportunity to demonstrate their ability to be creative problem solvers if the instructor is willing to accept what may seem like some unorthodox approaches to patient care. Carefully consider the plan the student has offered before offering a routine solution or before suggesting that the student look up the usual care as described in a standard nursing textbook.

There are a number of approaches to evaluating a written nursing care plan. The following is a suggested approach that has been used successfully by the author. It is assumed that the student has completed a nursing database and used the nursing process that included identification of nursing diagnoses. Because the care plan is only as good as the information available, it makes sense to begin the evaluation process by carefully reading the nursing database. Check for completeness of data in each category, writing comments as you proceed. For example, if the student has collected detailed information related to dietary habits including a listing of the typical intake for a 24 hour period, it would be appropriate to write a comment such as "nicely done" or "this information should help in your discharge planning". On the other hand, if the student has written "eats regular diet" it would be appropriate to write a comment such as "not enough data here to help you identify real or potential nutritional needs", or "it is important to find out what the patient means by 'regular diet' ".

While reading and commenting on the database identify potential nursing diagnoses and/or patient needs so that comparisons can be made between the nursing diagnoses made by the student and those made by the instructor. If there is consistency between the two lists then no further reading of the database is needed. If the student has failed to identify relevant diagnoses, you can either suggest the additional diagnoses based on your review of the database or you can suggest the student review the database for potential additional diagnoses. If the student has listed more diagnoses than you believe can be justified by the database, you can ask for supporting data since none seemed to exist in the database. If the students are expected to number the nursing diagnoses in order of priority, then asking them to justify their ranking can provide information about their ability to apply theory and principles in the practice setting. Blatant errors in judgment should be corrected, but be careful not to impose your own view when there might be an honest difference of opinion about the relative importance of one nursing diagnosis over another. It may be helpful to consider whether two experienced nurses might disagree about the ranking of the problems under consideration, and, if so, to explain your disagreement with the ranking but let the student's judgment stand. Remember, the goal is to help the student develop critical thinking, sound judgment, and independence.

One of the most common errors students make when developing a nursing care plan for a single nursing diagnosis relates to tailoring their plans to that diagnosis. For instance, if the nursing diagnosis relates to skin integrity then the subjective and objective data must relate to skin integrity, the analysis must be based on the subjective and objective data, the goal or objective must relate to skin integrity, each specific planned action must relate to skin integrity, each rationale must relate directly to a specific component of the plan, and the evaluation of the action must relate back to the effectiveness of the specific parts of the plan. Obviously the use of such a highly specific approach to the nursing care plan limits the number of nursing diagnoses or patient needs that can be addressed in one written assignment. Some instructors will set limits on the total number of diagnoses to be written up fully. Beginning level students may be asked to choose one physical and one psychosocial problem for complete work up; advanced students may be asked to choose a problem they have not encountered before.

Once the students have learned the basic process for writing detailed nursing care plans, they can use the same reasoning to write abbreviated care plans which can be added to the patient's "real" care plan. It is important for students to learn how to modify plans as patient conditions change. It is equally important that students learn to write nursing care plans which can be adapted to the variety of forms and formats they will encounter in different clinical agencies. The detailed care plans running many pages often required by schools of nursing can serve the purpose of helping students to demonstrate their understanding and skill in reasoning. However, the number of times a student should be required to produce such a care plan should be limited to the number of times it takes the student to show mastery of the process. After that it makes more sense to have the student use a more realistic, abbreviated form as a working document for each patient assignment. There are some instructors who argue that care plans are one of the best ways to make students analyze the reasons behind their nursing actions. I believe that a student's understanding of rationales can be just as easily determined by periodic, judicious questioning by the instructor or preceptor.

INTERPERSONAL PROCESS RECORDINGS

The interpersonal process recording (IPR) is a learning tool used most frequently in beginning level clinical courses and in clinical courses that focus on care of the mentally or emotionally disturbed. The form itself usually includes a description of the patient and the setting; an interaction goal; a format for recording and analyzing the verbatim interaction between the patient and the student; and a requirement that the student evaluate goal achievement. The format for recording the verbatim interaction is usually set up in a column format so that the student can record the conversation, describe nonverbal behaviors, identify interaction techniques used, give rationale for techniques used, and offer alternatives to ineffective responses. A typical format is shown in the figure below.

When evaluating interpersonal process recordings (IPR)

remember that what the student said during the interview is not as important as whether the student recognized that some of their comments were less than effective and whether the student offered reasonable alternative responses. Occasionally a student will have

Interpersonal Process Recording (IPR) Form				
Patient verbatim conversation	Nurse verbatim response	Communication technique	Analysis of effectiveness	Proposed alternate response

difficulty deciding between two or more similar communication techniques when labeling the nurse responses. Because effectiveness of communication is more important than the labels, it would be better to advise the student to analyze the effect of the response rather than to spend inordinate amounts of time on labeling.

Students should not be asked to offer alternatives to every part of their interactions with patients because this practice can inhibit their confidence in their ability to communicate with patients. If the responses they used were effective in achieving the goal of the interaction, then they should be accepted as they stand. As with care plans, there need to be guidelines for writing and evaluating IPR's. Inexperienced instructors are advised to seek the counsel of more experienced instructors to validate the ratings given to student written work.

MAINTAINING RECORDS OF STUDENT PERFORMANCE

Maintenance of records of student performance in the clinical setting is an important responsibilty of the clinical instructor. The records may take the form of anecdotal notes, audio-taped summaries of observations, or completion of a checklist of desired behaviors. Anecdotal notes provide great flexibility of content. In an anecdotal note the instructor records as objectively as possible a description of events and student behaviors.

Evaluative terms should be avoided. For example, the instructor would record, "bathed patient from head to foot, changed bathwater whenever it became soapy, used towel to cover wet skin areas, dried skin gently, maintained privacy for patient. Patient stated that she felt 'comfortable and clean' after the bath" This brief, but descriptive evaluation, is far more useful than an evaluative statement: "gave a good bath".

Another example of an objectively written anecdotal note might include the following statements: "Performed urinary bladder catheterization on female patient. Touched thumb of sterile glove with bare hand. Did not recognize error; changed gloves at direction of instructor. Cleansed perineal area with a circular motion rather than an anterior to posterior motion. Repeated cleansing of perineum at direction of instructor. Located meatus correctly. Maintained sterility of catheter and inserted it correctly. Allowed urine to drain by gravity. Removed catheter in one smooth motion. Did not remove gloves until all equipment had been gathered up. Put used equipment in wastebasket in patient's room but did not place articles in the readily available disposable wrapper from the catheter kit. Stated she 'felt nervous' and 'forgot all about the right direction for cleaning the perineum'." A less useful evaluative note might simply state "student used poor technique when doing cath." Well-written anecdotal notes provide the documentation needed to do meaningful evaluation of student clinical performance. They can be an invaluable aid in situations in which the student does not recognize his or her own weaknesses or errors. In reviewing a student's performance of a new procedure it is a good idea to first ask the student to do an informal self-evaluation before the faculty member provides her assessment.

Audio-taped summaries of student performance can take the place of written anecdotal notes. Either the written or verbal form of the anecdotal records should be constructed so that they are descriptive of the performance that was observed. Obviously, the more specific the notes the more useful they will be to both the instructor and the student. Regardless of whether taped or typed summaries are done they should be done in a manner that can be shared openly with the student. In fact, routine sharing of the instructor's observations with the student can be used as a mechanism to provide both positive feedback and constructive criticism.

Checklists of clinical or course objectives are another mechanism for maintaining records of student performance in clinical settings. Although checklists alone do not provide sufficient information to provide guidance to the student, those that include space for comments and examples often work quite well. As with anecdotal notes, the examples should describe rather than judge behavior. One of the major advantages of using clinical and course objectives as the basis for weekly clinical progress notes is they remind both students and instructors of the purposes of the clinical lab.

CONDUCTING EVALUATION CONFERENCES

Formal evaluation conferences are usually held at mid-term and at the end of the clinical course. The purpose of the mid-term conference is to discuss the student's progress toward meeting the course objectives. It is not possible to give a grade half way through a course. Nevertheless, the mid-term conference provides an opportunity for the student and the instructor to take some time out to identify strengths and weaknesses and to make plans to build on the strengths and overcome the weaknesses. The final conference of the term is really a summary conference. It is not a good time to be pointing up needs that could have been met during the term that just ended. That is the purpose of the mid-term meeting. If the instructor has provided ongoing feedback both verbally for performance and in writing for written assignments there should be no surprises at the final conference.

The format for the mid-term and end of term conferences is much the same. The conference time should be scheduled and private. The time may vary depending upon the student but an average conference can be completed in half an hour. Students should prepare by doing a self-evaluation using the same form the faculty will use. In most instances there will be great similarity between the student and faculty versions of the clinical performance evaluation. When there are great differences between the two versions it is often because the student lacks insight into her or his weaknesses. In these cases, it is essential to have available the documentation described above that illustrates the ways in which the student's behavior and performance fail to meet expectations.

Most authorities agree that clinical performance is best rated as satisfactory or unsatisfactory. To attempt a finer gradation of student performance in a clinical setting is neither fair nor defensible. The clinical setting includes too many variables over which neither the student nor the instructor has control. There are times when it is difficult to determine if the grade should be Satisfactory or Unsatisfactory let alone whether it should be "A", "B", or "C". It is the responsibility of the faculty members in a given clinical course to predetermine what behaviors constitute passing or failing. Most schools have definitions or descriptions of acceptable clinical performance. These expectations are likely to include adequate preparation for clinical, the ability to provide rationale for nursing actions and decisions, the provision of safe, effective care at a level of skill and competence that can be reasonably expected at a particular point in the program of study, the ability to recognize ones own abilities and limitations, and absolute honesty and integrity in patient care. If no such criteria are available the new instructor should seek consultation from senior faculty or the program director when student behavior or performance fails to meet stated learning or behavioral objectives. At the very least students must not pose a threat to patient comfort or safety. Some schools require that a second faculty opinion be sought in the event a student is failing the clinical component of a course. This policy is meant to protect both the student and the faculty from allegations of biased evaluation.

8
LEGAL CONSIDERATIONS OF CLINICAL TEACHING

Although you would prefer not to think about the possibility of lawsuit brought by students or by patients, such things do happen. Overconcern for legal actions might inhibit effective teaching and learning, but it is prudent to be aware of one's legally recognized responsibilities. This section contains a brief overview of legal considerations and responsibilities related to clinical teaching.

As a citizen, you are bound by the same laws that bind every other citizen and you are protected by the same laws that protect other citizens. Students, patients, and peers have the same responsibility and protection. As a professional, you are called to a higher standard of decision making in your area of expertise than would be the ordinary citizen. As a professional, you are bound by the additional laws and regulations that govern the practice of nursing. As a professional with advanced preparation, you are held to an even higher standard. The question faced by a clinical instructor in the clinical area becomes "what would a reasonable and prudent professional with similar education and experience do in the same circumstance?" It is well to remember that as an employee you are bound to follow the lawful demands of your employer.

The question of who is ultimately responsible for the care of the patient needs to be examined in light of where the patient is. If the patient is in a health care institution i.e., a hospital, nursing home, etc, the agency is ultimately responsible for the patient's care. This is one of the reasons that communication between staff and students and instructors is so important. The ultimate responsibility of the agency does not negate or reduce the responsibility of the individual caregiver whether that person is a staff nurse or a student. Since the 1970's, when young people were declared legal adults at age 18, students have been viewed as adults by the courts and are considered to be responsible for their own actions. This view does not remove responsibility from the instructor, but it does mean that the instructor is no longer solely responsible for the actions of the students. Students should not be assigned duties for which they are not prepared and they should not accept such assignments. Schools need to have mechanisms for recording when students are considered ready to perform certain procedures on actual patients and instructors need to be aware of the abilities and limitations of each student under their supervision.

The following are a series of areas of concern to consider during the planning, implementation, and evaluation of clinical learning activities.

1. Narcotics

Check the agency policy about signing out procedures and procedures for obtaining access to locked narcotic cupboards or dispensers. Many agencies require student signatures to be co-signed by the instructor who must be a registered nurse. Agencies frequently have co-signing policies when partial doses of narcotics need to be discarded. Such discarding usually has to be done in the presence of a witness.

2. Incident Reports

When a student makes an error or a patient has an accident or loses a possession, an incident report is filed in accordance with agency policy. The instructor usually co-signs the report. Some agencies send a copy of the incident to the school of nursing for filing with the student's record. Other agencies consider the incident report the property of the agency and no copies can be made.

3. Student Injuries

When a student has an accident resulting in personal injury or becomes ill and needs medical assistance the clinical agency will usually permit use of its emergency facilities or its staff clinic. The costs of such services are the responsibility of the student.

4. Contracts with Agencies

It is imperative that the school of nursing maintains contractual agreements with the clinical agencies used for student experiences. Most schools have a standard contract form that has been approved by the university's attorney. These contracts are general in nature and spell out the role of the faculty and the agency for providing learning experiences for students. The contracts are signed by the chief executive officer of the school and the clinical agency. Instructors should review the contract with the agencies they use for student clinical practice.

5. Evaluation of Student Performance

Evaluation of student performance is the prerogative and the responsibility of the instructor. As long as the instructor is fair in her or his application of the criteria for grading student performance and as long as the student is kept informed of her or his progress the courts are reluctant to interfere with the faculty member's right to be the final determiner of student grades. It is important for the instructor to be objective in the evaluation of student performance—particularly if the performance is unsatisfactory. It is equally important for the instructor to make the decision to fail a student who has not met the learning objectives. It is unfair to ones faculty peers, to patients, and to other students to permit a student who cannot perform safely and adequately in the clinical setting to continue to progress in the nursing program. Faculty have a significant responsibility as gate keepers of the profession. If you would not want a particular student to care for you or for some member of your family, why should the student be permitted to care for other vulnerable human beings?

Appendix A
Student Data Memo

Nightingale University
Clara Harlowe Barton School of Nursing

TO: Students in my clinical group
FROM:
SUBJECT: Things I want to know about you

Name:
Address:
Phone number: Cell phone:
Where was your most recent clinical practice?

Where did you have clinical practice last year?

Any health care experience prior to nursing school? Yes No.
If yes, explain.
Have you been employed in health care since enrolling in nursing school? Yes No. If yes, please explain.

Are you currently employed? Yes No.
    If yes, where?
    How many hours per week?
What kinds of clients and learning activities are you particularly interested in being assigned to during this clinical session?

Are there any types of clients or learning experiences that you particularly dread being assigned to during this clinical session? If so, please explain:

Please add any additional information that you think will assist me in understanding your learning needs and goals. (e.g., career goals, extra curricular responsibilities, etc.)

Appendix B
Agency Posting Form

Nightingale University
Clara Harlowe Barton School of Nursing
Clinical Assignments - First Year Students

Setting _____ Instructor _____
Date _____ Time _____

Clinical Study Objectives:

Room	Patient	Student	Special Data

Specific nursing care that all students may be expected to do for the patients they are assigned:

Appendix C
School Posting Form

Nightingale University
Clara Harlowe Barton School of Nursing
Clinical Assignments - First Year Students

Setting _____ Instructor _____
Date _____ Time _____

Clinical Study Objectives:

Student Name	Client Information			
	Age	Gender	Health Problem	Nursing Tasks

Specific nursing care that all students may be expected to do for the patients they are assigned:

Appendix D
Instructor Worksheet

Nightingale University
Clara Harlowe Barton School of Nursing
Clinical Study Assignments - First Year Students

Unit _____ Date(s) _____ Instructor _____

Student	Room	Patient Age Diagnosis	Diet I&O	TPR BP	Activity and Care Needs	Treatments and Meds.	Misc.

BIOGRAPHY

Helen Spustek O'Shea

Helen S. O'Shea received her diploma in nursing from the Martins Ferry Hospital of Nursing in 1958, her BSN and MS from the Ohio State University in 1961 and 1962, respectively, and her PhD from Georgia State University in 1980. She has been a faculty member and administrator at the Nell Hodgson Woodruff School of Nursing at Emory University since 1971. She retired as Professor Emerita from Emory in August, 2003. She currently directs and teaches in the post-master's summer teaching institute and continues to serve on the editorial board of the *Journal of Nursing Education*.

Dr. O'Shea primary areas of interest and expertise are undergraduate teaching, clinical instruction, and preparing graduate students for careers in nursing education.

Colophon

This book was composed and designed on Apple Macintosh computers using Microsoft Word for text, Adobe Photoshop and Apple Aperture for the cover and chapter photos, and Adobe Illustrator for the Chapter 4 figures. It was composited using Adobe InDesign. The body text is Garamond and the title typefaces are Arial. The book was published through lulu. com.

www.ingramcontent.com/pod-product-compliance
Ingram Content Group UK Ltd.
Pitfield, Milton Keynes, MK11 3LW, UK
UKHW041435180426
11947UKWH00007B/447